Inventions That CHANGED the WORLD

How the Steam Engine Changed the World

Jordan Johnson

Cavendish Square
New York

Published in 2019 by Cavendish Square Publishing, LLC
243 5th Avenue, Suite 136, New York, NY 10016

First Edition

Library of Congress Cataloging-in-Publication Data

Names: Johnson, Jordan (Children's author) author.
Title: How the steam engine changed the world / Jordan Johnson.
Description: First edition. | New York, NY : Cavendish Square Publishing, LLC, 2019. | Series: Inventions that changed the world | Includes bibliographical references and index. | Audience: Grades 5 to 8.
Identifiers: LCCN 2018019769 (print) | LCCN 2018020442 (ebook) | ISBN 9781502641182 (ebook) | ISBN 9781502641175 (library bound) | ISBN 9781502641168 (pbk.)
Subjects: LCSH: Steam-engines--History--Juvenile literature.
Classification: LCC TJ467 (ebook) | LCC TJ467 .J64 2019 (print) | DDC 621.109--dc23
LC record available at https://lccn.loc.gov/2018019769

Editorial Director: David McNamara
Editor: Kristen Susienka
Copy Editor: Alex Tessman
Associate Art Director: Alan Sliwinski
Production Coordinator: Karol Szymczuk
Photo Research: J8 Media

The photographs in this book are used by permission and through the courtesy of: Cover Bob Orsillo/Shutterstock.com; p. 4 Ronald Sumners/Shutterstock.com; p. 7 Bewickswan/Shutterstock.com; p. 8 Prisma/UIG/Getty Images; p. 9, 14, 15, 16, 22 SSPL/Getty Images; p. 10 Kodda/Shutterstock.com; p. 12 Uli Deck/Picture Alliance/dpa/AP Images; p. 18 DeAgostini/Getty Images; p. 28 George Rinhart/Corbis/Getty Images; p. 32 Chronicle/Alamy Stock Photo; p. 33 Vintage Power and Transport/Mark Sykes/Alamy Stock Photo; p. 34 Romaset/Shutterstock.com; p. 35 Bettmann/Getty Images; p. 37 Universal Images Group/Getty Images; p. 39 Albert Bierstadt/Yale University Art Gallery/Wikimedia Commons/File:Valley of the Yosemite by Albert Bierstadt 1868.jpeg/Public Domain; p. 44 Peter Titmuss/Shutterstock.com; p. 47 Studio BKK/Shutterstock.com; p. 49 Mark Boster/Los Angeles Times/Getty Images; p. 51 Three Lions/Getty Images; p. 52 Cyo Bo/Shutterstock.com; p. 54 Polaris/Newscom.

Printed in the United States of America

CONTENTS

Trains like this were
powered by steam engines.

CHAPTER ONE

The Engines of Progress

History is full of moments where humans learned to use the natural forces of the world to their advantage. They used rivers and streams to turn waterwheels; wind to turn windmills and push sails; and materials from the earth, like wood and stone, to make tools and homes. They used fire to keep warm, to cook food, and to make metal and glass. Each new technology made it possible for people to survive longer in an area, or move on and settle in another place.

For centuries, knowledge was used to make new technologies, but progress was slow. However, that changed when humans started using steam engines.

Steam itself wasn't a new idea. Anyone could make steam by boiling water in a pot or other vessel. But people found that if water was boiled in a sealed container, something interesting happened. With nowhere to go, the steam was trapped and pressure began to build up as even more steam was made. This pressure could be used and made to push objects. Eventually, the steam engine was born. The first steam engines used this pressure to push on moving parts. If the water was kept hot, a steam engine could repeat its task without stopping, as long as nothing broke.

Uses for Steam Engines

Not long after its invention, the steam engine was put to work. Steam engines could work for long periods of time, and that made them great for carrying out tasks that used to rely on muscles that could get tired or sore. Whether it was pumping water, grinding grain into flour, or moving heavy items along a railroad, steam engines could outperform muscle power. They were also used to replace energy sources that could be unreliable, like wind. Here are just a few ways humans have used steam engines over the centuries.

Pumping Water

Before the steam engine, people would get water either by gathering it from lakes, streams, or rivers, or by using pumps and wells to bring water up from below ground. Some people used aqueducts, or long manmade channels, to bring water sources closer to cities. All of these methods took lots of effort to prepare and use. Water is heavy, and carrying it over long distances is tiring. Using a mechanical pump by pushing on a lever also took lots of effort that could be better spent elsewhere. Steam-powered water pumps, however, could move enough water to support a much larger number of people with much less effort.

Transportation

Before steam engines, humans had very limited options for travel. If someone needed to get somewhere, most of the time they relied on muscle power. Riding a horse was certainly easier than walking or running, but horses had their limits. They got tired and sore, like humans do. They needed to be fed and maintained, even when not in use, and couldn't be used if they were sick or injured.

Steamships like this one used steam engines to turn large paddlewheels that would push on the water to move the ship forward.

DID YOU KNOW:

The world's fastest steam-powered train was the *Magnificent Mallard.* It had a top speed of 126 miles per hour (202.7 kilometers per hour).

By using steam engines to move wheels, people had a new option for transportation. The engines were used to power trains, which moved on rails and could cover greater distances in a shorter time than horses without getting tired.

The *Magnificent Mallard*, shown above, holds the world record for being the fastest steam locomotive.

Steam engines were used to speed up travel by sea as well. By spinning a fanlike underwater machine called a turbine, steam engines replaced the need for oars or sails. Not long after their first use, steam-powered ships were able to move twice as fast as some of the quickest sail-powered ships. Since they didn't have to rely on the wind to propel them, steamships could travel at top speed in any direction at sea. Their powerful engines also meant that steamships could be larger and could carry more people or cargo than sailing ships.

Manufacturing

One of the biggest benefits of the steam engine was its use in manufacturing. Steam engines were used in factories to complete tasks like shaping metal, quickly cutting wood, or creating textiles like fabric. Because they were powerful and could do the same task over and over, steam engines

Steam engines were very useful for improving production at lumber, textile, and grain mills like this.

STEAM AND ELECTRICITY

A modern power plant releases steam.

One of the ways that steam engines are still used today is to generate electricity. Power companies generate steam by burning fuels like coal or natural gas. Nuclear power plants also create steam by using radioactive materials to heat water. Steam from the heated water is used to turn the blades of a turbine. The turbine then produces electricity.

The electricity can be sent through power lines to customers in the area. In 2016, more than 80 percent of electricity in the United States was generated using methods like these. That means most homes, businesses, schools, and factories that use electricity are still indirectly powered by steam.

Other ways to generate electricity, like solar panels and wind turbines, are becoming more advanced and make up a larger portion of the United States' energy supply each year. Steam-powered electrical generators may be replaced by these other sources in the future, but for now they remain important parts of powering the world.

were able to make items more precisely, and were even used to help make more steam engines.

Steam engines were also used to gather resources. In mines, they were used to power machines that helped move ore and pump water out of tunnels. On farms, they powered machines used to plow fields, harvest crops, and haul products from place to place. Because steam engines did these tasks efficiently, people could have larger farms and produce more crops to feed more people.

Leading to a Revolution

Since steam engines made travel, manufacturing, and gathering resources faster, more people could live in the same area. This led to the rise of larger cities and towns. Because more people were living closer together with less worry about how they would get food, water, or shelter, a period of rapid technological improvement was possible. This era, powered by steam engines and other technologies invented at that time, became known as the Industrial Revolution.

The steam engine helped lead to a cycle of progress. Each new technology it helped create enabled people to take on problems they couldn't tackle before. The steam engine also helped to spread new ideas and art through faster travel. Although steam engines are not used today the same way as in the past, their impact on human productivity remains revolutionary.

This modern steam turbine generator burns coal to make power.

CHAPTER TWO

The Story of Steam

It's easy to think that the history of steam engines started during the age of locomotives, smoky factories, and heavy machines. However, the steam engine has a long history, beginning in ancient times. Each generation sought to harness steam's power and make it into a functional resource. The steam engine's past is filled with moments of small improvements. The technology was reshaped many times by inventors, allowing it to become more and more powerful.

Ancient Steam Engines

Although the 1700s and 1800s are when steam engines became more advanced, people had long known about steam. After all, people have been boiling water for thousands of years. However, not many people thought

This is a replica of an aeolipile, with pipes connecting the basin to the ball and pipes to release steam.

to use the power of steam to do something. The very first recorded steam engine comes from the first century CE. An engineer named Heron of Alexandria (also sometimes called Hero of Alexandria) made a device called the aeolipile.

The aeolipile, also known as a Hero engine, was a container of water with a heavy lid. The lid had two pipes leading up to a ball-shaped chamber. The ball also had two or more spouts attached to the sides. As the water was heated over a cauldron and boiled, the steam created would produce pressure and be forced up the pipes, into the ball, and out the spouts. The steam coming out of the spouts would cause the ball to spin. Back then, this device was considered a curiosity or a toy rather than a serious device.

Although they weren't steam engines, steam was used in other ways in the centuries following the aeolipile. In early twelfth-century France, a church organ powered by steam was recorded by William of Malmesbury. In the fifteenth century, inventor and painter Leonardo da Vinci tried to design a cannon that could use steam pressure to launch a cannonball. It didn't work very well.

A Renaissance for Steam

In the 1600s, multiple inventors from around Europe began to work with steam. One of the first major steps

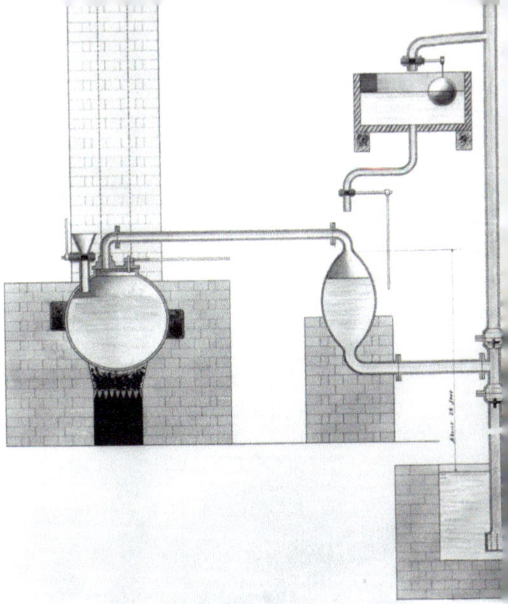

This diagram shows the parts of Thomas Savery's steam pump.

was the use of steam to create a vacuum. This led to the use of steam pressure to move water. In 1601, Giambattista della Porta, an Italian inventor, used vacuum pressure created by steam to push water through a fountain. In 1606, Jerónimo de Ayanz y Beaumont, a Spanish engineer, patented the first steam-powered water pump for mining.

Since steam pumps could be fueled by coal, they became particularly popular in coal mining operations. By using coal to power the pumps, miners were able to reach deeper coal deposits and provide coal to pumps at other mines. This led to more metals being available, which played an important role in building and improving new steam engines.

In 1698, Thomas Savery created a steam pump that was more powerful and less expensive than other pumps. His design, called "the miner's friend," could also work at greater depths. However, the machine itself became known for exploding if it overheated. This opened the door for an inventor who could make a safer steam-powered pump.

Building Up Steam

The 1700s is the century when the steam engine really began to take form. In 1705, an English engineer named

THOMAS NEWCOMEN

Thomas Newcomen was born in 1664 in England. From an early age, Newcomen took an interest in working with metal, and he learned the skills of a blacksmith. As an ironmonger, some of his major clients were workers in English tin mines, who needed metal tools to do their jobs.

Newcomen learned through his clients that flooding in the tin mines was a big problem. The deeper the mines reached, the more water would fill in. This water would need to be pumped out.

Newcomen set to work designing a way to remove water from the mines that was inexpensive, easy, and safe. Despite the fact that Thomas Savery had already made a steam-powered pump, Newcomen didn't give up. To get around the fact that Savery owned a patent, or claim, on his steam pump, Newcomen became Savery's business partner.

Newcomen worked on designing his engine until Savery's patent expired. From there, he set about building and installing the engines that made him famous.

This is an atmospheric engine. The first one was invented by Thomas Newcomen.

Thomas Newcomen designed one of the first practical steam engines. His device was called the atmospheric engine. It used steam pressure to push on a circular plate called a piston. The piston was in a metal tube called a cylinder. When the piston was pushed all the way up, the machine sprayed water into the steam to cool it down. The cooling steam created a vacuum, which sucked the piston back down to its starting position. Using this method, the machine could repeat the up-and-down motion and act as a more powerful pump than Thomas Savery's invention. In 1712, Newcomen and Savery worked together to build the first working version of Newcomen's design.

In 1713, a boy named Humphrey Potter made a simple change to the steam engine that made a big difference. Newcomen's engine required a person to work the machine by opening and closing different valves at the right time. Potter made the machine operate itself by tying one of the machine's working parts to the lever for the valve. All that was needed to keep the machine going was to keep the boiler hot.

In 1718, an engineer named Henry Beighton created a safer and more reliable version of Potter's idea. In 1733, Newcomen's patent on his engine expired, and engineers in England were free to make their own versions of Newcomen's engine.

In 1755, the first steam engine was installed in the United States by Josiah Hornblower. Using Newcomen parts brought over from England, Hornblower's engine was used to pump water from a copper mine in New Jersey.

Over the next few decades, steam engines similar to Newcomen's design were modified and improved. They were used in textile mills and factories, not just for pumping water from mines.

James Watt: Heating Up and Cooling Off

There are two very important things a steam engine needs to work. First, it needs pressure to push its piston. Second, it needs vacuum pressure to pull the piston back into place. Newcomen's engine created vacuum pressure by spraying water into the piston's cylinder. However, this engine was still pretty slow. Even after the engine was made able to operate independently, its top speed was around fifteen strokes per minute. That meant the engine could do a complete cycle of motion once every four seconds. Part of this was because the spray of water cooled the cylinder too much. This meant more steam was needed not only to push the piston but to heat up the cylinder too. Every time the cycle started, the machine lost some heat and needed more fuel to burn and heat up again.

This is a model of James Watt's steam engine.

In 1765, an engineer from Scotland named James Watt envisioned and made a major improvement to the steam engine. His improved engine was different in three main ways.

First, he made a cooling system that didn't cool down the cylinder as much. Instead of spraying water inside the cylinder, steam was pushed into a separate area to cool. The hot water left over could still be used elsewhere. Second, he used valves to put steam in different places during the engine's motion. This let him use steam to push the piston both ways instead of relying only on a vacuum to pull it down. Vacuum pressure still played a part, but not as much. When steam cools down enough to condense back into water, it can still be quite hot. So, third, Watt designed a way to take that hot water and put it back into the boiler. From there, it could be heated up again and remade into steam. This meant the boiler needed less fuel to keep the whole machine running. All of these improvements made the steam engine more powerful and more efficient.

Further Improvements

Watt continued to improve upon his design over the course of his career. One important change Watt made was figuring out how to turn the pumping or churning motion of a steam engine into torque, or the energy to turn a wheel or gear. He did this by using something called a flywheel. A flywheel is a heavy wheel attached to the steam engine. Without a flywheel, a wheel pushed by a steam engine would vibrate because of the back-and-forth motion of the piston. A heavy flywheel fixed this by being able to spin forward when the piston pushed on it, but not backward when the piston moved back. Because the wheel was heavy, it was hard to start moving but easy to keep going. The smooth spinning motion of a flywheel made it easier to attach gears to the engine to power machines.

Another important invention of Watt's was the governor. Invented in 1788, the governor controlled an engine's speed without affecting its efficiency much. It consisted of two or more weights at the end of levers. A gear connected the governor to the flywheel, allowing it to spin. As the weights spun, the circular motion would raise the levers to which they were attached. The levers connected to a valve that would slow the engine's water supply if it got too fast. When the engine slowed down, the governor would release a little. Eventually, the engine would reach a balance between the speed of the engine and the weight on the governor's levers, allowing the engine to maintain speed.

The use of a governor was an important step for Watt's engine to replace Newcomen's pump. The governor also made it possible to use Watt's engine for tasks where high speed wasn't needed, like in grain milling.

High-Pressure Steam

Just as Watt improved the Newcomen engine and Newcomen improved the Savery pump, a new inventor found a way to change Watt's steam engine for the better. Around the turn of the nineteenth century, an inventor named Richard Trevithick created a high-pressure steam engine.

By this point, most engineers knew that steam pressure was the driving force behind the steam engine. However, the danger of too much pressure was also well known. Ever since the Savery pump was made infamous for exploding, pressure became both an ally and an enemy to a steam engineer. Part of the problem was that in the days of Savery's pumps and Newcomen's engines, boilers, cylinders, and other important parts for a steam engine were made the old-fashioned way, by hammering metal into shape. Pipes and smaller parts were made by casting molten metal in a mold. However, by the 1770s, Trevithick discovered that while most steam engines could withstand pressure in their cylinders, boilers were often about as simple as a pot or a kettle. Not much pressure built up there. Trevithick found that by making a boiler large and strong enough to hold lots of pressure, large amounts of steam pressure could be stored. This steam pressure could be used to push on both sides of a piston, like in Watt's engine, but without the need for vacuum pressure. Since it didn't need vacuum pressure, it also didn't need to pump water to condense the steam back into water. This also meant the cylinder could be smaller. Having smaller engines that were still efficient and powerful was an important step toward making steam engines useful for more tasks.

Since most of these parts were metal, making them smaller meant that steam engines became much lighter and more portable, or able to be moved from place to

Trevithick's early locomotive design was at first intended to work on roads rather than rails.

place. A smaller, lighter, more powerful steam engine quickly led to the idea of using a steam engine to move itself. That feat was first accomplished in 1769, when Nicholas-Joseph Cugnot developed the first steam-powered vehicle. Although it was not very practical, it affirmed such a task was possible. Decades later, in 1804, Trevithick introduced the world to the first steam-powered railway vehicle. He put a steam engine on a cart and used the engine's flywheel to pull the cart. The device became known as a "locomotive," a term that meant it could move itself. However, because Trevithick's design used only one flywheel to do all the work, it wasn't very fast or balanced.

The person who really helped the railway industry, and by extension the steam engine, succeed was George Stephenson. Stephenson was a mechanic who enjoyed learning about and improving mining technologies. Eventually he became familiar with and perfected the locomotive design. He introduced the first public passenger locomotive, *Active*, in 1825. Today he is known as the father of the railway. Without him, railways may not have existed and steam engines may not have been as popular.

Steam at Sea

Steam engines changed water travel as well. In 1807, an American engineer named Robert Fulton built a riverboat that used a steam engine to turn a paddle wheel. The steam-powered boat was strong enough to travel along the Hudson River from New York City to Albany, New York. This was important because traveling against the current of a river wasn't so easy at the time. While sails could help a vessel get part of the way upriver, oars or man and animal power on shore would have to help the rest of the time. Steamboats, on the other hand, could use their engines to get upriver much more easily. Workers on a steamboat could save energy by not having to row, and the shipping company saved money by not having to hire workers to pull the boat upriver.

Riverboats like Fulton's were good for traveling along rivers and coasts, but getting across the ocean was another matter. Crossing the Atlantic Ocean from Europe to the United States was a long and dangerous journey. With good winds and weather, the journey on a sailing ship took about six weeks. After the steam engine became popular, many ships were built with both steam engines and sails. This allowed sailors to use wind when they could and steam when the wind or weather was poor.

The Industrial Revolution

The first few centuries of steam power saw slow progress at first with later leaps and bounds made by inventors and engineers. However, during the 1800s, a great number of improvements were made to the steam engine. These changes took place within a period of rapid technological advancement called the Industrial Revolution. It spanned from the 1700s to the mid 1800s.

DID YOU KNOW:

The first ship to cross the Atlantic with the help of a steam engine was the SS *Savannah* in 1819. The journey took just under one month.

Although it had been in use for decades grinding grain into flour or pumping water from mines, steam power was used more widely during the latter part of the Industrial Revolution. It especially helped and affected people who lived in cities. Advancements in steam engine technology and a wider availability of coal helped spur demand for steam-operated factories and machines.

Factories that used steam engines could make more products faster than before. A steam engine could move a saw to cut wood faster than people could. It could move a heavy grinding wheel to grind large amounts of grain into flour. Factories also used steam engines to repeat the same task over and over to create nearly identical items that could be sold at the same price. This included things like clothes, tools, and food containers.

Dangers of Steam Engines

Steam engines could be very dangerous. They combined extreme heat, high pressure, and moving parts to work. Someone using or working near a steam engine could face a variety of hazards from the machine. One of the first dangers came from its main source of fuel: coal.

When mining, shoveling, or working with coal, small bits of coal were often knocked off as coal dust. This coal dust was light enough to float on the air for a while, or be blown around by wind. Coal miners and workers whose jobs were to use coal to keep a steam engine's fire going often breathed in this dust accidentally. Coal dust is harmful to humans if enough gets into their lungs, and causes severe breathing problems. To make matters worse, a burning piece of coal could also cause a fire. This was particularly dangerous on ships at sea. To reduce the chance of fires, people used great caution when working with coal. To avoid coal dust, some workers covered their noses and mouths with cloth.

Another danger of steam engines was heat. In order to boil water, the engine needed its boiler to be heated. Most boilers and other steam engine parts were made of metal. This was because metal was a strong material and could absorb and hold heat easily. This meant almost any part of a running steam engine could be very hot. Accidentally touching the steam engine in the wrong place could give someone a nasty burn. Also, because steam itself was quite hot, it could burn someone. If a steam engine broke and sprayed steam, it could cause serious harm to anyone caught in the spray. Avoiding touching the machine was enough to avoid most burns. To prevent dangerous steam leaks, engineers and smiths took care to make parts as sturdy and precise as possible.

One of the most spectacular dangers of steam engines was their ability to explode. If a steam engine built up too much pressure, the metal parts containing the pressure could break. This would cause extremely hot steam and broken pieces of metal to fly outward with incredible force. Even though Thomas Savery's pump didn't use any moving parts, it was made famous for exploding under too much pressure.

To reduce the chance of explosions, engineers would make thicker, sturdier pieces that could safely contain steam pressure. One important invention to avoid explosions was the safety valve. Invented in 1718 by Jean Desaguliers, its job was to release pressure to prevent explosions. Desaguliers first used his invention to make Savery pumps safer to use.

Patents

Explosions, burns, and steam were dangerous, but they were not the only obstacles engineers and inventors faced when considering the engine's creation. There was also a problem of patents. A patent is a license that allows someone to claim all the profits off an invention they make for a certain amount of time. They're an important part of rewarding a great inventor, but they can still be a problem for others who want to create similar inventions. One example of this was when the Savery pump was popular. Thomas Savery had a patent on his invention that gave him the right to profit from all other industrial uses of steam. This meant that Thomas Newcomen, who had improved Savery's pump, couldn't try to sell his own steam engine design. Newcomen had some options, though. He could wait until Savery's patent expired, but that could take many years, or he could overcome the patent restriction by becoming partners with Savery. He chose to become partners. Savery agreed.

Since they were working together, Savery let Newcomen continue his work to profit them both. Once Savery's patent expired, Newcomen got a patent of his own for his steam engine. Newcomen's patent then made it hard for other engineers to submit their own designs. Any new designs couldn't be too similar to Newcomen's.

Another example of patents posing an obstacle to inventors came in 1781. Jonathan Hornblower had invented his own variation on the steam engine that was more powerful and more efficient than James Watt's engine. However, Watt owned the patent on steam engines at the time, and Hornblower's design was very similar to Watt's. Because of this, Watt was able to stop Hornblower from getting his own patent.

Today, patents are often less of an obstacle for engineers because the way patents are made and issued has changed. Patents on designs today tend to be more specific. This prevents people from patenting all similar technologies, like Thomas Savery did, limiting other inventors like Newcomen. Additionally, patents often only apply in certain nations. Patents issued in England wouldn't necessarily stop an American engineer from making their own version of a similar design, for example. However, these developments don't prevent people from continuing to abuse the patent system. In some cases, individuals or companies, known as "patent trolls," will acquire as many patents as they can, including patents on overly broad ideas. To make money, the patent trolls then sue lots of companies and inventors for allegedly infringing on their patents.

This steam locomotive is being used to move large bins to carry ore or coal.

CHAPTER THREE

The Impact of Steam Engines

Steam engines were great tools that enabled humans to do things they couldn't do before, such as travel farther and faster, perform tasks quicker, and make more goods in a shorter amount of time. Because of these advancements, the steam engine made a huge impact on society in a few centuries. Overall, it moved people from a more agrarian lifestyle into a modern era of machines and new technologies, paving the way for other advancements in following generations.

Mines, Metals, and Miscellaneous Materials

Mining coal, metals, and other minerals from the earth isn't an easy task. Today, modern miners have new tools and techniques to make mining easier and safer. However, before the steam engine, working as a miner was a very difficult and dangerous job. Cave-ins, poisonous and explosive gases, and other dangers were common events workers experienced in mines. One of the biggest problems with deep mining tunnels was flooding. Natural underground water sources and any rainfall would cause water to fill the deeper parts of the mines. This meant miners had to get water out of the mines in order to reach the resources below. Ore and coal deposits usually run very deep, meaning most of the ore or coal would likely be below the water level. If miners couldn't get to it, the mine might close and they would not be able to work there.

Before Savery's pump, miners worked pumps manually or used buckets tied to ropes pulled by animals. These methods were tiring, slow, and inefficient. But if the water wasn't removed, miners could only get to the resources above the water level. Both Savery and Newcomen set out to solve this problem by inventing different ways to pump water. Savery's steam-powered pump was the first, but Newcomen's later, improved version was more effective.

Both pump designs allowed miners to remove more resources from the mines, making them available to inventors and engineers. The abundance of resources like coal and oil was an important part of the rapid progress in the following century.

For a long time, steam engines mostly saw use in coal and ore mines. They were popular because they were cheaper, more powerful, and safer than Savery's pumps. By the 1730s, more than one hundred Newcomen engines were in mines all over England. Owners of mining operations liked that the engines were effective, easy to maintain, and cheaper than other methods. When Newcomen's patent expired in 1733, use of steam engines spread even more quickly.

Without the restriction of Newcomen's patent, engineers could build and improve on his design. By the early 1770s, more than six hundred new engines were in use. However, it was Watt's design that was more popular outside of mines. Watt's engines had a big advantage in being able to provide rotating energy instead of just back-and-forth motion. By connecting the engine to other devices like cranks, gears, and flywheels, the engine could provide rotation for all sorts of tasks. Lumber mills could use them to spin circular saws to cut wood. Grain mills could use them to turn heavy grinding stones to make flour. Watt's engines' flexibility made them useful for more tasks other than pumping water from mines.

Turning Steam into Power

By the 1880s, steam engines had become very popular. They were used in mills and factories to make and move items with ever-increasing speed and efficiency. However, they were still facing some old problems. Steam engines that relied on pistons still couldn't make a smooth and even amount of energy. A piston was either pushing and contributing energy or it wasn't. That changed in 1884 when an inventor named Charles Parsons invented the steam turbine.

BEASTS OF BURDEN

For centuries, horses were used to pull boats and other vessels upriver.

The steam engine replaced a lot of tasks that humans relied on animals like horses or mules to do. For example, riverboats no longer needed the help of mules or horses to pull them along as they docked at a port. Likewise, donkeys were no longer needed to turn a mill. Mail delivery also changed. Mail was carried by train or steamboat rather than by animals and stagecoaches. Steam engines didn't need to be fed, cleaned, or looked after as much as working animals did, and only used fuel when they were actively working. They could also accomplish tasks much more quickly. This made them the favored choice for factories, shipping, and transport industries for much of the Industrial Revolution and after.

Despite steam engines replacing them as the main modes of transport and strength, horses and mules were still used when a few people or items needed to travel a short distance or for specialized tasks like in cattle ranching or farming.

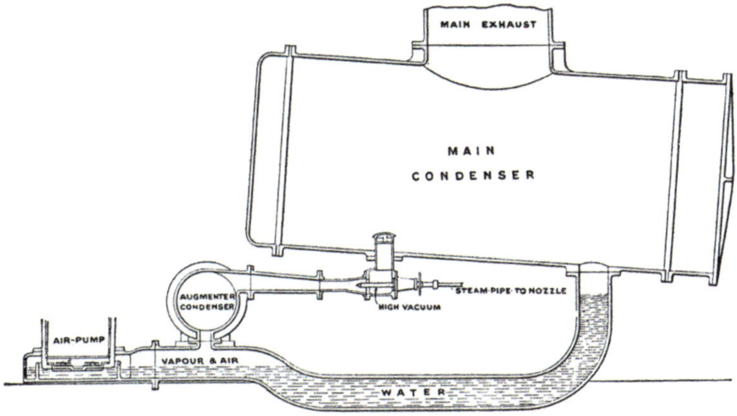

This diagram shows the parts of a steam turbine condenser, which helps the steam turbine change steam into water.

Unlike previous steam engines that used pistons, a steam turbine used moving steam to push on a series of fanlike blades. As the blades spun, they moved at a steady rate as long as the amount of steam pushing on the blades stayed the same. Instead of using the turbine to move machines in factories and mills, Parsons connected his turbine to a dynamo, a device that used a spinning motion to create electrical energy. By connecting the turbine and dynamo, Parsons had created the first steam-powered electric generator.

Electricity was a major improvement for society. Before its widespread use, the only way for people to get light at night was to burn something, such as lamp oil or wood in a fireplace. Streets at night were lit with oil lamps or by burning gases pumped through pipes beneath the streets. Electricity was helpful because it didn't cause smoke. Electric generators could provide a constant stream of energy and light, while other methods would dim when fuel was almost used up.

Turbine generators like these use the flow of water to generate electricity.

Parsons's steam turbine generator was a huge leap forward for humanity. With access to a steady stream of electrical energy, lots of new technologies were invented. Homes and businesses with access to electricity could use lightbulbs to see indoors. All sorts of devices powered by electricity were invented, built, and sold to people who could use them to make life easier. With every new device, people got more free time. With more free time, people could spend it getting an education or working on other tasks to make life easier. Some of those people would create new inventions that also improved the quality of life, and so on in a cycle of progress. Today, other generators have been invented, but many use turbines similar to Parsons's.

Smoke and Pollution

In order to heat water in a steam engine, it needs to burn fuel. The most popular fuel for steam engines was coal because it was cheap and there was lots of it. By the 1800s, millions of tons of coal had been mined in the

This woman uses a cloth to filter polluted air in London in 1953.

United States and England to use as fuel for steam engines.

However, burning coal sent clouds of smoke into the atmosphere, and it didn't take long before the air began to be polluted. Coal smoke changed from an annoying part of daily life to something dangerous. People started to have a hard time breathing in areas with lots of coal smoke. Natural weather patterns would combine with coal smoke and other fumes from factories to create smog. Although both the United States and England passed laws to try to reduce coal smoke, things got worse. In 1873, London suffered the first of what became known as "killer fogs." These fogs combined normal fog with dangerous air pollution that had built up over the years. It was dangerous enough that many people died as a result of the bad air. Even worse, after laws aimed at reducing smoke and other air pollution were passed, enough pollution remained for more fogs to occur over time. The worst recorded killer fog occurred in 1952, nearly eighty years after the first. It killed more than ten thousand people and caused hundreds of thousands more to suffer from severe breathing conditions.

Disease

Pollution wasn't the only new danger brought on by steam engines. Because steam engines allowed more people to live closer together, diseases also became a problem. Some diseases, however, were made worse because of the use of steam engines.

In the early 1800s, people didn't quite understand that diseases were caused by germs like viruses or bacteria. Therefore, people weren't trying to stop bacteria or viruses from spreading. People also didn't understand that human waste could carry disease, so cities would dump sewage into the same river that it pumped water from to drink.

One way this became a problem was with a disease called cholera. In the 1830s, 1840s, and 1850s, England and especially the city of London faced cholera epidemics.

To supply water to the city, steam-powered water pumps were stationed along the river. These pumps would supply water to nearby wells in the city. Unfortunately, a few of these pumps were nearby a sewer outlet that dumped waste into the river. This meant that some of the water pumped back into the city was dirty and could carry diseases like cholera.

The problem was finally solved by a doctor named John Snow. He realized that only people who drank from certain wells and water pumps in the city got sick. In 1854, he famously removed the handle from the pump on Broad Street in London to stop the spread of disease. He also traced the problem back to the steam-powered pumping station that was bringing dirty water into that part of the city. Once people realized that dirty water could spread disease, they quickly developed ways to prevent it from happening again. Today, cities all over

This steam-powered riverboat uses a paddlewheel to travel upriver in Australia.

the world use sewer systems to keep wastewater separate from drinking water so people don't get sick.

Steam Transportation

One of the largest impacts the steam engine had on humanity was its contribution to transportation. Upon their introduction, steam-powered locomotives, riverboats, and steamships quickly outpaced other methods of transport. Because humans could travel or carry things faster and farther with steam engines, people quickly saw the benefits of such machines. Any town near a railroad had better access to anything that could be carried by train. Rivers and large bodies of water were less of an obstacle because steamboats and steamships could cross them quicker than by sail. People, resources, news, and mail were all easier to get and stay connected to because steam engines could help them make the journey.

One of the most iconic forms of transportation brought about by the steam engine was the locomotive. A locomotive was a large vehicle that used a large steam engine to pull itself. Locomotive engines were often connected to a series of cars connected in a train. During the 1860s, locomotives had an average speed of 60 miles per hour (96 kmh). That was nearly twice as fast as a running horse. Because trains were so fast and could carry heavy loads, they became very popular for travel and deliveries. In 1862, the United States tasked two railroad companies with building a line that would connect the western railways with the eastern railways. In 1869, the Union Pacific and Central Pacific railroad companies completed the transcontinental railroad. This meant that someone could travel from New York to California in less than a week. In earlier years, the same journey by horse and wagon would have taken several months.

For much of the 1800s and into the 1900s, railroads became a vital part of life in the United States, and many people relied upon them. Almost anything that could fit in a train car could be carried by rail, and resources of all kinds were transported across the country. Railroads would remain the dominant mode of transportation until automobiles became popular in the 1900s.

Processing Information

One of the most impressive things about humans is our ability to share complex ideas. Language and writing are two of our most valuable tools. However, the steam engine also played a big role in helping people communicate and spread ideas.

THE RISE OF ENVIRONMENTALISM

Albert Bierstadt's painting of Yosemite helped raise support for conservation.

Because people could travel faster and farther using steam engines, more people got to see new places. These new places inspired people and encouraged some to discuss their visits. By the 1860s, several artists, writers, and journalists had visited a beautiful natural spot called Yosemite Valley. The natural beauty of Yosemite's untouched land captivated them. Several famous paintings, articles, and books were made, highlighting the importance of natural places like Yosemite. This led to the rise of environmentalism, the idea that untouched areas of nature should be preserved.

With better access to natural beauty, people started wanting to protect nature. They began to work to pass laws to permanently protect places like these. Through their efforts, famous forests like Yosemite were protected from loggers. Famous landscapes like Yellowstone became national parks protected by the United States government. As more laws were passed and new beautiful places were discovered, more and more people considered protecting the environment very important. Without the steam engine, perhaps programs like the National Park Service would not be in place today, working to preserve the United States' natural areas.

Although printing presses had been around for centuries, the amount of text they could print was based on their speed. A few trained workers could print 250 pages of text in an hour using hand-cranked and muscle-powered printing presses. When the steam engine was put to use as a printing machine in 1814, that changed completely. A steam-powered printing press could print more than 1,000 pages an hour, even printing on both sides. By 1830, it could do 4,000 pages per hour.

Not only was the steam engine being used to print on paper, but it was being used to make paper. In the early 1800s, steam engines were being used in paper mills to grind wood into pulp and roll it into paper. With more paper and faster printing, more books, newspapers, and other writing supplies could be used to share information and ideas anywhere they could be delivered.

DID YOU KNOW:

One of the first laws passed that tried to limit air pollution was the Smoke Nuisance and Abatement Act in London, England, in 1853.

A Changing Climate

Around the 1850s and 1860s, people were noticing a negative side to steam-powered innovations. The smoke that had poured for decades from factories, ships, and trains without any regulations had especially impacted the environment. Polluted air was affecting the weather, causing things like acid rain to damage stone buildings and monuments. Polluted water was spreading disease and causing erosion along coasts and rivers. There were also signs of an idea that would become known as the greenhouse effect.

In 1863, a scientist named John Tyndall gave a famous lecture on the greenhouse effect. He explained that burning lots of coal releases gases that are better at keeping heat in Earth's atmosphere. The greenhouse effect is a process that warms Earth's surface, like a greenhouse warms plants. Gases like carbon dioxide and methane help trap the heat. One major way to make a lot of carbon dioxide is by burning fossil fuels like coal, petroleum, and natural gas. Over the course of the steam engine's history, millions of tons of coal had been burned as a way of fueling homes and machinery. Scientists had been taking temperature readings and studying weather since the 1700s. This meant that when changes did start happening, people were able to observe and record them. This change could have wide-reaching effects on the global climate.

Today, multiple nations are working together through the United Nations to reduce the use of fossil fuels. In doing so, they hope to slow or reverse the toll that centuries of burning fossil fuels has taken on the atmosphere.

The Decline of Coal

In the late 1800s, coal miners were growing more upset. Mining was hard and dangerous work, and workers felt they were being treated unfairly. Their pay wasn't very good, and the shifts didn't leave workers much time with family. This led miners to start going on strike, or refusing to work until their needs were met. Some of the most famous strikes were at Pennsylvania coal mines, which were a major source of coal for the United States. The strikes had a significant effect on people who relied on coal for heat and fuel. With the workers on strike, there was a major coal shortage. This meant that the price of coal for fuel increased. This also led to people seeking out a new fuel source that was cheaper and more plentiful. Although the strikes ended with coal miners getting better pay and safer working conditions, the damage was done. Coal was on its way out as the world's favorite fuel.

The fuel that replaced coal was petroleum oil, a dark liquid that could be burned directly or refined into other fuels. It was cheaper, easier to gather, and more efficient than coal in many ways.

One major invention that led to the dominance of petroleum fuels was the internal combustion engine. Rather than burn fuel to heat water to push on pistons, the internal combustion engine burned fuel inside the cylinder directly. This eliminated the boilers needed to fuel steam-powered engines, and allowed the engine to be smaller, lighter, and more powerful.

The rise in popularity of petroleum fuels posed a problem for engineers who used steam engines on trains and in factories. Without coal, the machines wouldn't work the same. However, at this time electrical energy was also becoming more popular, and one of the most effective ways to make it was with steam turbines. It was a simple fix for engineers to convert turbine generators from using coal to using other fuels. Since most of the work happened after the steam was made, the only thing that needed to change was converting to a liquid fuel source instead of solid coal.

Another fuel that saw more use in turbine generators was natural gas. In the 1800s, natural gas was mostly burned for light. In the 1900s, it began to be used to run steam turbine generators and as a source of heat for homes and ovens. Natural gas is still used for these things today.

Because steam turbines are able to use almost any source of heat to work, they're very adaptable. Being able to operate on new fuels and sources of heat keeps steam turbines useful.

This geothermal power plant operates in New Zealand.

CHAPTER FOUR

Steam Today

In the twenty-first century, the days where steam engines dominated have passed. In fact, steam engines themselves are rarely if ever used. Many countries favor cars and other vehicles with internal combustion engines for their power, versatility, and the easy access to fuel. Factories, mills, and other manufacturers use electricity and other types of energy to get the job done faster and more efficiently than a steam engine. However, the legacy of the steam engine is a lasting one, and the world will see the effects of it for decades to come.

Steam Without Smoke

The primary use of steam today is in power plants to generate electricity. These power plants use heat to create steam. The steam turns turbines to create power. In some

DID YOU KNOW:

As of 2016, the United States generates the most geothermal energy in the world. However, geothermal energy makes up less than 0.5 percent of the amount of energy used in the United States.

cases, fossil fuels like coal, petroleum, and natural gas are burned to create heat. In the case of nuclear power plants, radiation from elements like uranium is used to heat water and make steam.

There is one form of steam power that takes no fuel, makes no smoke, and can provide renewable energy for homes and businesses. All it takes is the natural heat produced underground by our planet. The energy created from this heat is called geothermal energy.

By drilling holes deep into the ground, humans can gain access to the heat made by Earth's hotter inner layers. By using these heat sources, humans can boil water, make steam, and turn turbines to make electricity.

According to the US Energy Information Administration, there are three main ways geothermal energy is used to make electricity. They are called dry steam, flash steam, and binary steam. Each method of geothermal steam has advantages and drawbacks.

Dry steam methods use naturally occurring sources of extremely hot water, like geysers, to make steam. This method is easy to use because all it takes is a way to capture the steam and direct it at a turbine. However, relying on certain natural formations, it is impossible for

This diagram shows how geothermal plants make power.

all places to produce power using this method. In the United States, the only places that could be used for dry steam are in Yellowstone National Park and The Geysers, an area in California. National laws forbid the use of Yellowstone for this purpose, so The Geysers is the only source of dry steam in the United States. According to the Geothermal Energy Association, California's geothermal plants account for about 80 percent of the United States' geothermal energy.

The flash steam method uses deeper pockets of hot water underground to make steam. The underground pockets of water are under lots of pressure. The pressure prevents the water from boiling into steam. By drilling down into the pockets of water, humans can allow the water to be pushed up a pipe using its own pressure. As the water goes up, it is able to boil and make steam. The steam can then be used to make electricity through a turbine. When the steam cools down, it becomes water again, which is then dropped back down underground to be heated.

Binary steam works a lot like flash steam, except the water heated underground is used to boil a second liquid other than water. This other liquid, called a "working fluid," boils and works like steam to turn a turbine. Like in flash steam, any cooled water is injected back down into the ground to be heated again. What makes it different is that sealing off hot water from the working fluid keeps fumes like sulfur from escaping and possibly polluting the surrounding area.

Biofuels

One problem facing the use of fossil fuels to make steam is that they are becoming scarce. Because they take so long to naturally form and are being used frequently in vehicles, factories, and other activities, they are running out. The other problem is the harm burning fossil fuels does to the environment. Renewable electricity sources like solar power and geothermal energy can help replace commonly used fuels. However, fossil fuels are hard to replace completely because they're easy to store. Coal, natural gas, and petroleum can be contained and moved around without losing much effectiveness. Current renewable technologies have a harder time doing this.

One possible way to solve these problems is by using different fuel types. A new fuel type would need to be plentiful and cheap. It would also need to cause less harm to the planet and be easy to store. Finally, it would need to be about as efficient as fossil fuels. One popular field of research to answer these needs is biofuels.

Biofuels are made from plants grown today. Like fossil fuels, biofuels are often burned. They cause less harm to the environment than fossil fuels. In 2012, the US Department of Energy measured that the biofuel

This algae farm in San Diego, California, grows algae to make biofuel.

ethanol released between 19 percent and 49 percent fewer greenhouse gases than gasoline. Currently, biofuels are most popularly added to gasoline and other fuels like diesel or jet fuel to make vehicles or planes run. Adding biofuels to fossil fuels helps reduce the greenhouse gas emissions of these sources. A buildup of greenhouse gases is leading to warmer temperatures in the atmosphere. This is called global warming. Adding biofuels to already existing fuel sources may sound good, but it doesn't entirely eliminate the emission of greenhouse gases into the air. Therefore, scientists are studying other, more promising fuels to use.

Today, biofuels are made from a variety of sources. Crops like corn and sugarcane are used to create ethanol, a biofuel often mixed with gasoline. Soybean and canola oil are used too. They can either replace or be mixed with diesel fuel. Another source of biofuels is algae. Algae are microscopic organisms that use sunlight to make their energy and food. With careful farming and processing, algae can produce lots of chemicals needed to make

ELON MUSK AND STORING ELECTRICITY

Elon Musk is a South African businessman and scientist known for founding multiple technology companies. He has led and funded projects to make affordable electric cars, further space exploration and spaceflight technology, and tackle multiple problems with environmental impacts. He has also turned his attention to renewable energies such as solar and wind.

One downside of wind and solar generators is that without wind or sun, they're producing no power. Solar panels and wind turbines don't store energy. They only produce it.

In 2017, Elon Musk touted a new project. It took the form of a giant battery pack that could store large amounts of electricity generated by wind and solar power. His focus was on Southern Australia, which has some of the highest electricity bills in the world.

The deal to build the battery pack was officially announced in September 2017. Musk had also promised to construct it in "100 days or it is free." He and his company, Tesla, met that goal, and the battery pack began charging on December 1, 2017. Within weeks, the pack was proving its worth, stepping in when a coal plant tripped on December 14.

Like the rise of the steam engine powered the Industrial Revolution, rises in technologies like Musk's battery pack may lead to another revolution, but one of renewable resources rather than fossil fuels.

biofuel. One advantage of algae biofuels over corn- or soybean-based fuels is how little space they require. Algae can make fuel faster and in less space than it takes for corn to grow. One acre (0.4 hectares) of land growing corn can produce 420 gallons (1,590 liters) of biofuel. One acre (0.4 ha) dedicated to algae biofuel can produce 5,000 gallons (18,927 L). This means there is less worry about how much land to use for food crops and how much to use for fuel crops.

Transportation Without Steam

The Ford Model T was a popular early automobile.

Steam-driven locomotives are no longer commonplace on railways around the world. Although enthusiasts still celebrate the era of the steam train, other engine types have all but replaced steam engines for travel. However, some methods of travel today still use technologies originally paired with steam engines.

Internal combustion engines get their name from how they work. Rather than burn fuel to heat external parts, fuel is burned inside the engine. First, fuel is pumped into the cylinder, then the fuel is ignited. The expanding gas of the burning fuel pushes the piston, providing kinetic energy. Because there is less need to keep a constant level of heat, internal combustion engines can work without needing time to warm up. Today, internal combustion engines are found in most cars, trucks, and other road vehicles. However, even the internal combustion engine might be replaced by a different engine type. For example, cars powered by electricity are becoming more popular as they compete more with gasoline-powered vehicles.

Trains might not be powered by steam, but travel by rail is still commonplace around the world. Most modern trains use internal combustion engines, but some rail vehicles use electricity to move. Some trains use electric motors connected to a source of power through the railroad itself. The railroad carries power to the vehicles on it, and the vehicles use that power to work electric motors.

Other railways use electromagnets, sources of magnetic energy powered by electricity, to move. These electromagnetic trains, sometimes called maglev trains, have several advantages over other train types. The biggest is that they're very fast. They're fast because the magnets keep the train hovering just above the rail, so there isn't any friction between the train and the rail. They also have fewer moving parts because the magnetic rails are used to propel the train and also to slow it down. They're also quieter and have fewer emissions than trains that use fossil fuels like diesel.

Maglev trains like this use magnets for fast, smooth travel.

Flying Turbines

One technology we have today that can trace its design back to steam engines is the turbo-fan engine. Turbo-fan engines are seen on many passenger planes and other aircraft. A turbo-fan engine works a lot like a turbine generator. A series of rotating fanlike blades spin and move air in one side and out the other. Unlike steam engines, turbo-fan engines use an internal fuel source to move the fan blades.

Turbo fans aren't the only plane engine to use turbines. Jet engines use turbines to pull air into the engine as well. Jet engines also burn fuel inside the engine, using the hot expanding gases to push the aircraft forward and to turn even more turbines.

New Engines for the Future

Today, steam engines are mostly only used for making electricity. For other former uses of steam engines, such as in vehicles, more advanced engine types are used. The most popular engine for vehicles is the internal combustion engine. However, new engine types may one day replace it. The main way engineers are hoping to replace current engine technologies is by making an engine that is just as effective, without damaging the environment by emitting greenhouse gases.

Most engines today require a fuel to be burned in order to work. However, engineers and inventors have come up with new engine types that don't need to burn fuel at all. So far, the most common is the electric motor.

Semitrucks like this one built by Elon Musk's company Tesla can use electricity alone to move, saving fuel.

An electric motor uses electricity to turn its wheels and move forward. Because no fuel is burned, electric-powered vehicles are quieter and release no greenhouse gases when used. However, electric vehicles rely on batteries, and battery technology has its limitations. For example, cold temperatures can cause batteries to lose effectiveness. However, as new and better battery technology is created, electric vehicles are becoming more and more popular.

One engine type that sidesteps this issue is the hydrogen fuel cell engine. This engine type uses hydrogen gas to create electricity without burning anything. Hydrogen is stored in a tank in the vehicle, just like any other fuel today. The hydrogen is pumped into the fuel cell where it reacts with oxygen and creates electrical energy and a little water vapor. The water vapor can then be captured and stored in a separate tank to be separated back into hydrogen and oxygen using electricity. With a fuel cell instead of a conventional electric battery, cold temperatures are less of a problem. Fuel cell engines also don't emit any greenhouse gases.

One other engine type today uses radiation to generate electricity, just like a nuclear power plant. The main vehicles that use nuclear-powered engines are submarines and large ships designed to break through icy water. The *Curiosity* rover that was sent to Mars in 2012 uses a small nuclear reactor to generate electricity and heat to keep itself running on the planet's cold surface. However, most people consider nuclear energy to be too dangerous to use in personal vehicles today.

The Legacy of Steam

The very first steam engines were little more than a few metal parts and a pot of boiling water. Today, they are so much more. Even though the use of steam engines eventually led to climate change, they also led to new technologies that make them an important part of human history. From the humble water pumps of the 1600s to the powerful generators in use today, the technologies and advancements we celebrate are due in part to the steam engine. While the use of fossil fuels may now endanger natural life, they also once enabled humans to see it. Today, humanity enjoys a great number of tools and technologies that make life easier and enable us to experience more of the world. Very few of them would have been possible without the steam engine.

GLOSSARY

aeolipile A device invented by Heron of Alexandria that could use steam to move.

boiler A container of water that is heated to create and channel steam.

climate change The change in global climate caused by the emission of greenhouse gases into Earth's atmosphere.

cylinder A can-shaped device used to work with a piston to create a back-and-forth or up-and-down motion.

dynamo A device that converts kinetic energy into electrical energy.

engineer Someone who works with and around engines as a profession.

flywheel A device used to conserve rotating kinetic energy from things like a steam engine.

fossil fuel A source of fuel formed by the decay of organisms through natural processes on Earth over the course of millions of years.

governor A device designed to limit the speed of a steam engine or another device.

inventor A person who creates a new object or device to fulfill a need.

locomotive A vehicle capable of moving itself.

patent A license or official document granting an inventor exclusive rights to profit off of their invention.

piston A moving part within a cylinder designed to harness pressure and convert it into kinetic energy.

pressure A type of force exerted on something.

steam power The use of steam engines to perform work.

turbine A series of fanlike blades used to convert moving steam or other fluid into rotation energy.

United Nations An organization made up of many countries around the world that decide on important global issues.

valve A device used to allow or stop the flow of steam or other fluids through a pipe or tube.

FURTHER INFORMATION

Books

Burton, Anthony. *The Steam Engine: Pride of Britain.* Stroud, UK: The History Press, 2012.

Drury, George H. *Guide to North American Steam Locomotives.* Waukesha, WI: Kalmbach Publishing Company, 2015.

Helfand, Lewis. *The Industrial Revolution.* Campfire Graphic Novels. New Delhi, India: Campfire Publishing, 2017.

Kras, Sara Louise. *The Steam Engine.* Transforming Power of Technology. Langhorne, PA: Chelsea House Publishing, 2003.

Websites

Ancient Origins: The Ancient Invention of the Steam Engine by the Hero of Alexandria

http://www.ancient-origins.net/ancient-technology/ancient-invention-steam engine-hero-alexandria-001467

This article covers the aeolipile, the spinning steam ball that was the first recorded machine to use steam power.

Catskill Archive: The Evolution of the American Locomotive

http://www.catskillarchive.com/rrextra/absa1.html

This article details the history of locomotives and steam trains in America.

Interesting Engineering: James Watt, Father of the Modern Steam Engine

https://interestingengineering.com/james-watt-father-of-the-modern-steam engine

This article explores the career and achievements of James Watt in his work with steam engines.

Videos

Coal, Steam, and the Industrial Revolution: Crash Course World History

https://www.youtube.com/watch?v=zhL5DCizj5c

This video provides a brief and interesting perspective of the Industrial Revolution and its effect on the world.

Steam Engine: How Does It Work?

https://www.youtube.com/watch?v=fsXpaPSVasQ

This video shows how steam engines work at various stages in the history of the engine.

SELECTED BIBLIOGRAPHY

"Algal Biofuels." Department of Energy. Accessed March 27, 2018. https://www.energy.gov/eere/bioenergy/algal-biofuels.

Baidawi, Adam. "Australia Powers Up the World's Biggest Battery—Courtesy of Elon Musk." *New York Times*. November 30, 2017. https://www.nytimes.com/2017/11/30/world/australia/elon-musk-south-australia-battery.html.

Bain, David Haward. *Empire Express: Building the First Transcontinental Railroad*. New York: Penguin Books, 2000.

"Better Than Corn? Algae Set to Beat Out Other Biofuel Feedstocks." Accessed April 16, 2018. http://www.worldwatch.org/node/5391.

Coulson, Michael. *History of Mining: The Events, Technology and People Involved in the Industry That Forged the Modern World*. Hampshire, UK: Harriman House, 2012.

"Geothermal Explained." US Energy Information Administration. November 28, 2017. https://www.eia.gov/energyexplained/index.cfm?page=geothermal_environment.

"History—James Watt." BBC. Accessed March 27, 2018. http://www.bbc.co.uk/history/historic_figures/watt_james.shtml.

"Industrial History: The History of the Steam Engine." ThomasNet. Accessed March 27, 2018. https://www.thomasnet.com/articles/custom-manufacturing-fabricating/steam engine-history.

International Correspondence Schools. *Steam Engineers Handbook: A Convenient Reference Book for All Persons Interested in Steam Boilers, Steam Engines, Steam Turbines, and the Auxiliary Appliances and Machinery of Power Plants.* London, UK: Forgotten Books, 2015.

National Renewable Energy Laboratory. Accessed March 27, 2018. https://www.nrel.gov.

Rose, Joshua. *Modern Steam Engines: An Elementary Treatise Upon the Steam Engine, Written in Plain Language: For Use in the Workshop as Well as in the Drawing Office.* Seaside, OR: Watchmaker Publishing, 2012.

"Thomas Newcomen." Encyclopedia Britannica. February 24, 2011. https://www.britannica.com/biography/Thomas-Newcomen.

Wilda, Hermann, and Charles Salter. *Steam Turbines: Theory & Construction.* La Vergne, TN: Merchant Books, 2007.

INDEX

Page numbers in **boldface** are illustrations.

ABOUT THE AUTHOR

Jordan Johnson is a writer and history enthusiast from Wisconsin. During his free time as a kid, he spent many hours researching vehicles of all types, including those from the age of steam. He has written on subjects ranging from airplanes to spaceflight to the history of environmentalism.